LIFTOFF! SPACE EXPLORATION

GREAT BALL OF FIRE

STUDYING THE SUN

I0794188

BY JONATHAN AND MARIEL BARD

Gareth Stevens
PUBLISHING

Please visit our website, www.garethstevens.com. For a free color catalog of all our high-quality books, call toll free 1-800-542-2595 or fax 1-877-542-2596.

Cataloging-in-Publication Data

Names: Bard, Jonathan. | Bard. Mariel.
Title: Great ball of fire: studying the sun / Jonathan and Mariel Bard.
Description: New York : Gareth Stevens Publishing, 2021. | Series: Liftoff! space exploration | Includes glossary and index.
Identifiers: ISBN 9781538258866 (pbk.) | ISBN 9781538258880 (library bound) | ISBN 9781538258873 (6 pack)
Subjects: LCSH: Sun–Juvenile literature.
Classification: LCC QB521.5 B37 2021 | DDC 523.7–dc23

First Edition

Published in 2021 by
Gareth Stevens Publishing
111 East 14th Street, Suite 349
New York, NY 10003

Copyright © 2021 Gareth Stevens Publishing

Designer: Katelyn E. Reynolds
Editor: Abby Badach Doyle

Photo credits: Cover, p. 1 Lukasz Pawel Szczepanski/Shutterstock.com; cover, pp. 1–32 (series background) David M. Schrader/Shutterstock.com; p. 5 Naeblys/Shutterstock.com; p. 7 (inset) Smithsonian Institution (https://www.flickr.com/people/25053835@N03) from United States/File Upload Bot (Magnus Manske)/ Wikipedia.org; p. 7 (main) Omnidoom 999/Wikipedia.org; p. 9 Dmitry Rukhlenko/Shutterstock.com; p. 11 (inset) Andreas Cellarius/File Upload Bot (Magnus Manske)/Wikipedia.org; p. 11 (main) steve estvanik/ Shutterstock.com; p. 13 Universal History Archive/Getty Images; p. 15 (main) NASA/MSFC/David Higginbotham; p. 15 (inset) Erik Simonsen/ Photographer's Choice / Getty Images Plus; pp. 17, 21 (inset), 29 NASA Goddard; p. 19 (main) NASA/Carla Thomas; p. 19 (inset) NASA/Scott Kelly; p. 21 (main) ESA&NASA/SOHO; p. 23 NASA; p. 25 NASA's Goddard Space Flight Center; p. 27 NASA/Johns Hopkins APL/Steve Gribben.

All rights reserved. No part of this book may be reproduced in any form without permission in writing from the publisher, except by a reviewer.

Printed in the United States of America

Some of the images in this book illustrate individuals who are models. The depictions do not imply actual situations or events.

CPSIA compliance information: Batch #CS20GS: For further information contact Gareth Stevens, New York, New York at 1-800-542-2595.

Find us on

CONTENTS

WORDS IN THE GLOSSARY APPEAR IN **BOLD** TYPE THE FIRST TIME THEY ARE USED IN THE TEXT.

THE SUN

Since the earliest civilizations, humans have been interested in the sun. What is it about this bright, burning gas ball that fascinates us so much? Since ancient times, we have known the sun is powerful. Some cultures even worshipped the sun. It's easy to understand why: The sun provides so much for us on Earth, from the sunlight that helps plants grow to the seasons we experience. Even when the sun has set for the day, we feel its absence in the night.

However, what do we really know about the sun? Over thousands of years, we've made observations and calculations to try to get to know our sun better. As science and technology have advanced, so has our knowledge of the bright star that brings order to our lives.

INTERSTELLAR INFORMATION

WITHOUT THE SUN, LIFE WOULD NOT EXIST ON EARTH! OUR PLANET WOULD BE TOO COLD FOR ANYTHING TO SURVIVE.

YOU WOULD NEED 109 EARTHS LINED UP END TO END TO SPAN THE DIAMETER OF THE SUN, AND ABOUT 1.3 MILLION EARTHS WOULD FIT INSIDE IT.

HOW BIG IS THE SUN?

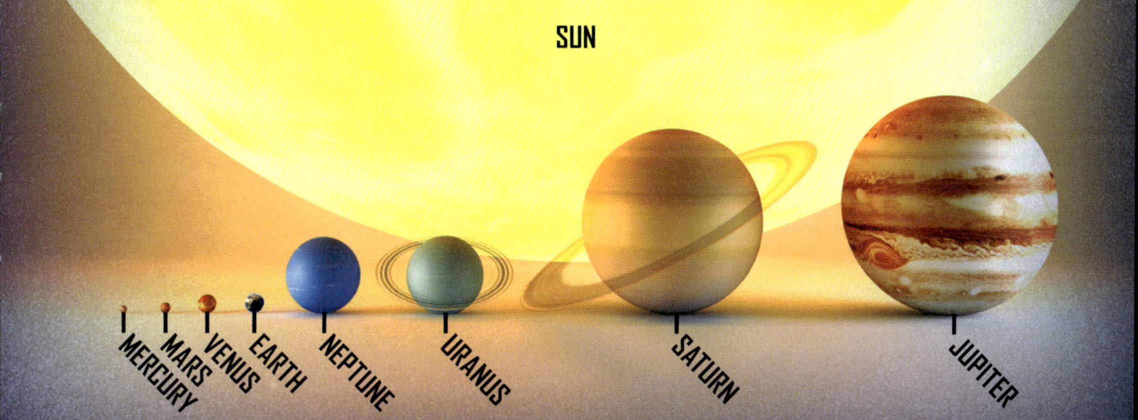

THE SUN IS MASSIVE!

The sun is the biggest object in our solar system, and it's a near-perfect sphere, or ball shape. It's also the most massive object in our solar system. In fact, between 99.8 and 99.9 percent of the mass found in our entire solar system is the sun. That means everything else makes up less than 0.02 percent of the solar system's total mass. That 0.02 percent includes all the planets, their moons, and any comets and asteroids too.

SUNS ARE STARS

Our sun is a star, just like the ones twinkling in the night sky. However, our sun is much closer to us than other stars. Stars have differences in size, age, and temperature. Scientists label stars based on these qualities.

In 1901, astronomer Annie Jump Cannon worked on the Harvard classification system. This uses temperature to place a star into a lettered category: O, B, A, F, G, K, and M. Type O stars burn the hottest and appear blue, while type M burn much cooler and appear reddish orange.

In 1943, W. W. Morgan, Philip Childs Keenan, and Edith Marie Kellman developed the Morgan-Keenan (or MK) system, which further grouped stars based on how much energy they give off. These two systems are what scientists use today.

INTERSTELLAR INFORMATION

OUR SUN IS A G2V STAR, ACCORDING TO THE MK SYSTEM. THE "G" MEANS IT APPEARS MOSTLY YELLOW, THE "2" REPRESENTS HOW HOT IT IS (ON A 0 TO 9 SCALE, WITH 0 BEING THE HOTTEST), AND THE "V" TELLS US IT'S A DWARF **MAIN SEQUENCE** STAR.

ANNIE JUMP CANNON IMPROVED THE OLD SYSTEM BY ORGANIZING STARS IN A SIMPLER WAY. HER WORK SHOWED OTHER SCIENTISTS THAT WOMEN CAN BE JUST AS BRILLIANT IN ASTRONOMY AS MEN.

STAR TYPES IN THE HARVARD SYSTEM

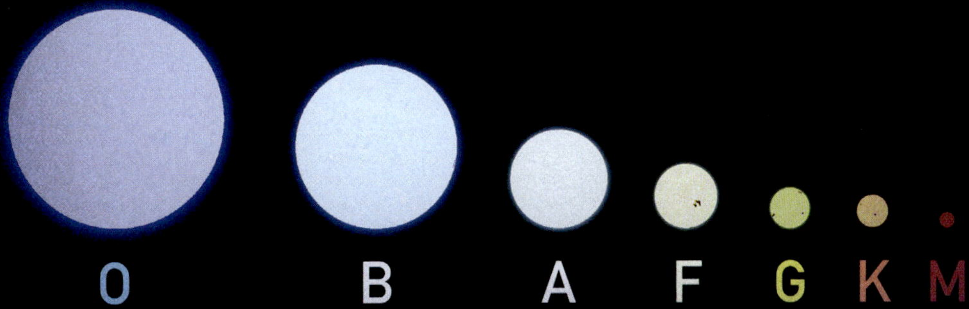

STARS IN SEQUENCE

A star's life starts as a cloud of the **elements** hydrogen and helium. Once that condenses into a ball, 90 percent of stars become a main sequence star. This sequence, or order of how things happen, represents where a star is in its hydrogen-burning cycle. As it ages and burns more of its hydrogen, it becomes brighter and hotter. Eventually though, main sequence stars burn up all the hydrogen in their core and drop out of the main sequence.

ANCIENT ASTRONOMERS

For thousands of years, humans have watched the sun rise and set, day after day. We have built our lives around this cycle. Before the invention of various scientific instruments, ancient cultures made observations to try to understand the sun and its effect on Earth.

Some civilizations, including the Maya, Aztecs, ancient Egyptians, and ancient Romans, considered the sun so important that they worshipped it as a god. Those cultures and others also tracked the sun's movement across the sky and celebrated it with holidays, feasts, and even monuments. You might know the Maya pyramids as tourist attractions. Originally, these structures were built to align with the sun on important days, such as the winter and summer solstices and the vernal and autumnal (or spring and fall) equinoxes.

INTERSTELLAR INFORMATION

ON THE SHORTEST DAY OF THE YEAR, ICELAND'S CAPITAL CITY, REYKJAVIK, HAS ONLY 4 HOURS AND 8 MINUTES BETWEEN SUNRISE AND SUNSET!

THIS MAYA PYRAMID IN CHICHEN ITZA WAS BUILT IN A WAY THAT IT CASTS A SERPENT-LIKE SHADOW ON THE SPRING AND FALL EQUINOXES.

THE LONG AND SHORT OF IT

An equinox occurs when the sun is exactly above the equator. It results in equal amounts of day and night. Earth has one equinox in spring and another in fall. A solstice occurs on the longest day of the year and again on the shortest day. The Northern Hemisphere experiences the most daylight on the summer solstice in June and the least daylight on the winter solstice in December. In the Southern Hemisphere, the seasons are opposite.

GEOCENTRIC MODEL

Early astronomers built a model of Earth in relation to the sun, moon, planets, and stars they could see in the sky. This early attempt put Earth at the center of it all. Everything else, including the sun, was believed to revolve around Earth. This is the geocentric model. "Geo-" comes from the Greek word "ge" meaning earth, and "centric" means at the center.

Astronomers believed this model was correct based on what they had observed at the time without the help of advanced scientific instruments or even telescopes. In the second century CE, the Greek philosopher, mathematician, and astronomer Ptolemy published these observations in his work *Almagest*. The geocentric model became widely accepted and would stay that way for the next 1,500 years.

INTERSTELLAR INFORMATION

THE GEOCENTRIC MODEL MADE SENSE FOR A LONG TIME BECAUSE PEOPLE COULDN'T IMAGINE EARTH CONSTANTLY MOVING. THEY DIDN'T UNDERSTAND GRAVITY (GALILEO GALILEI AND ISAAC NEWTON WEREN'T AROUND JUST YET). THEY THOUGHT EVERYONE WOULD FLY OFF INTO SPACE IF EARTH WAS SPINNING!

DUTCH CARTOGRAPHER ANDREAS CELLARIUS CREATED THIS SCENOGRAPHY OF THE GEOCENTRIC MODEL IN 1660. SCENOGRAPHIES ARE PAINTINGS THAT USE ARTISTIC **PERSPECTIVE** TO REPRESENT SCENES, IN THIS CASE THE SOLAR SYSTEM.

A WORKING, SPINNING MODEL

Scientists use models to help explain their observations. A model can be as simple as a drawing or as detailed as a working reproduction built to scale. Astronomers use an orrery, which is a mechanical version of the solar system as seen from the outside. (The one shown here has the sun at its center.) The more modern orrery is believed to have been invented in the early 1700s by George Graham and named after the Earl of Orrery. Versions that came before were known as planetariums.

HELIOCENTRIC MODEL

The idea that Earth was the center of the universe was not popular with everyone. Just as Ptolemy was finishing is geocentric model, another astronomer and philosopher, Aristarchus of Samos, proposed that Earth revolved around the sun—not the other way around. This model is called the heliocentric model, from the Greek term "helio" meaning sun. Over the years, several other astronomers proposed the heliocentric model, including Indian astronomer Aryabhata in 499 CE.

A major shift away from the geocentric model occurred when mathematician Nicolaus Copernicus developed a math-based model suggesting the sun is at the center of our solar system. Copernicus's 1543 work (published just before he died) is titled *On the Revolutions of the Heavenly Spheres* and it laid the foundation of the modern heliocentric model.

INTERSTELLAR INFORMATION

ARISTARCHUS OF SAMOS DEVELOPED HELIOCENTRIC IDEAS ABOUT 1,400 YEARS BEFORE COPERNICUS AND WITH MUCH EARLIER TECHNOLOGY. RESEARCHERS ONLY HAVE REFERENCES TO HIS WORK, BUT IT'S CLEAR HE WAS AHEAD OF HIS TIME.

IT TOOK MORE THAN 100 YEARS FOR COPERNICUS'S THEORY TO BE WIDELY ACCEPTED. THE CATHOLIC CHURCH FINALLY AGREED WITH THE HELIOCENTRIC MODEL IN 1822.

IT'S A REVOLUTION!

At first, the Catholic Church accepted Copernicus's theory, but the Protestant Church did not. Later, the Catholic Church also banned his teachings. Therefore, supporting Copernicus's theory was dangerous. In fact, Italian scientist Giordano Bruno was burned at the stake in 1600 for teaching this and other ideas not approved by the Church. Slowly, improvements were made to the original theory by astronomers such as Galileo Galilei, Tycho Brahe, and Johannes Kepler (who worked with Brahe).

THE FIRST TELESCOPES

A major advance in astronomy occurred with the invention of the telescope. The first telescopes, developed in 1608 in the Netherlands, were able to magnify objects by three times using a series of curved, clear glass called lenses. Italian astronomer Galileo Galilei heard about this invention and improved on its design. His version improved the magnifying power more than 20 times, making distant objects such as the moon and sun much more visible.

Modern-day telescopes use mirrors, which are able to collect and amplify light much better than glass. Mirrors are also easier to polish, which creates a clearer image. Many mirrors can be used to collect and focus any available light in a dark sky, allowing astronomers to clearly study even faint objects very far away.

INTERSTELLAR INFORMATION

GALILEO DID MOST OF HIS OBSERVATIONS OF THE SUN DURING SUNRISE AND SUNSET, WHEN THE LIGHT OF THE SUN WASN'T AS DANGEROUS. HE EVENTUALLY SWITCHED TO THE PROJECTION METHOD, DEVELOPED BY HIS STUDENT BENEDETTO CASTELLI.

TELESCOPE TECHNOLOGY HAS COME A LONG WAY, FROM THE SIMPLE GLASS USED IN THE 1600s TO THE 18 HEXAGONAL MIRRORS IN THE JAMES WEBB SPACE TELESCOPE SHOWN HERE.

SHIELD YOUR EYES!

It's very important to not look directly at the sun without special eye protection. The bright light can permanently damage your eyes. You can also use a method called projection. First, point a telescope or lens toward the sun. Then direct the other end of your telescope toward a sheet of paper or screen. The sunlight passes through the telescope, which then projects onto the surface of the paper or screen.

SPOTS ON THE SUN

Looking through his telescope, Galileo noticed something peculiar: The sun had dark spots. Now called sunspots, these are areas of cooler temperatures on the sun's surface that appear darker than the rest of the sun. Galileo noticed that these sunspots would move and were irregularly shaped. This meant that the sun wasn't the flawless heavenly body the Church wanted people to believe.

Galileo took notes, but he kept quiet until German astronomer Christoph Scheiner observed and reported on sunspots in 1611. However, Scheiner claimed that the spots were just **satellites**, like moons or planets, around the sun. When Galileo heard this, he published his findings stating that the spots were actually on the sun and that the sun rotates, which explained how the spots could travel.

INTERSTELLAR INFORMATION

THE FIRST WRITTEN RECORD OF A SUNSPOT COMES FROM CHINESE ASTRONOMERS IN 28 BCE WHO OBSERVED A BLACK SPOT AT THE CENTER OF THE RISING SUN.

SCIENTISTS USE TELESCOPES TO TRACK AND MEASURE SUNSPOTS. ONE SUNSPOT CAN BE MANY TIMES THE SIZE OF EARTH, MEASURING TENS OF THOUSANDS OF MILES ACROSS.

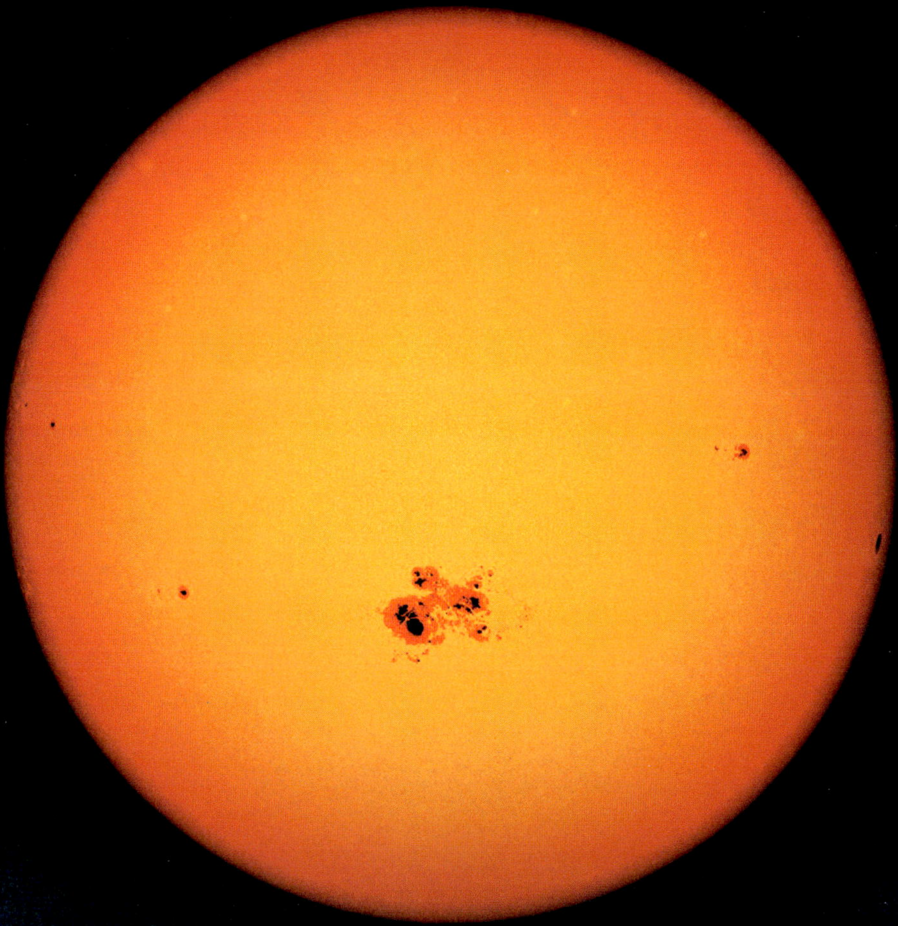

THE MAUNDER MINIMUM

Shortly after the discovery of sunspots, scientists began tracking how many spots happened in a year. Over time, they noticed trends. In the period shortly after Galileo, the number of sunspots dropped dramatically for about 70 years. This period became known as the Maunder Minimum, named after astronomers Annie Russell Maunder and Edward Walter Maunder. In acknowledgement of their scientific contribution, the Maunders have a crater on the moon named after them!

SOLAR FLARES

In 1859, two scientists witnessed a sudden, **intense** burst of light from the sun. Later, telegraphs stopped working. Strange, colorful lights appeared in the sky. Some people thought the world was ending! Today, we know this was a solar flare. A solar flare is a sudden explosion of energy released from the sun near sunspots.

One type, called a coronal mass ejection (CME), sends a huge amount of **radiation** into space at an extremely fast speed. Sometimes, that energy is directed toward Earth. When CMEs head for Earth, radio and electrical equipment can be affected by the sudden increase in energy. One such CME hit Earth in 1989. It caused an energy surge and a massive power outage in Quebec, Canada, as well as parts of the United States.

INTERSTELLAR INFORMATION

THE STORM THAT FOLLOWED THE 1859 SOLAR FLARE IS KNOWN AS THE CARRINGTON EVENT. IT'S NAMED AFTER RICHARD CARRINGTON, ONE ASTRONOMER WHO OBSERVED IT. THIS FLARE HAD THE ENERGY OF 10 BILLION ATOMIC BOMBS.

IN A SOLAR ECLIPSE, THE MOON CROSSES BETWEEN THE SUN AND EARTH. TOTAL ECLIPSES MAKE IT EASIER TO SEE THE SUN'S CORONA, OR THE OUTER LAYER WHERE SOLAR FLARES OCCUR.

AURORA BOREALIS AS SEEN FROM THE INTERNATIONAL SPACE STATION ▶

NORTHERN AND SOUTHERN LIGHTS

Sometimes, beautiful colors light up the night sky. These dancing lights are caused when the sun's radiation collides with Earth's magnetic field. The electrons and protons sent from the sun crash into oxygen and nitrogen atoms in our atmosphere to create light. In the Northern Hemisphere, this is called the aurora borealis. In the Southern Hemisphere, it is called the aurora australis. During solar flares or coronal mass ejections, the auroras get stronger and more visible.

SOLAR WINDS

The sun's core produces a huge amount of energy. It releases that energy as light and heat. The corona is so hot that the sun's gravity cannot hold onto particles there. Tiny charged particles, mostly protons and electrons, escape the sun's gravity and travel through space at very high speeds. These stream out in the form of plasma, or charged gas. This stream of plasma and particles on the move is called the solar wind.

Solar winds are continually streaming toward Earth; our satellites and probes, or unmanned spacecraft; and even astronauts in space. Still, there is much we don't know about how they work. Scientists are very interested in how solar winds form. NASA and other space agencies around the world have launched many satellites to study solar winds.

INTERSTELLAR INFORMATION

ACCORDING TO NASA, SOLAR WIND IS MOST LIKELY RESPONSIBLE FOR MARS LOSING ITS ATMOSPHERE. MARS DID NOT HAVE A STRONG ENOUGH MAGNETIC SHIELD TO PROTECT IT, CAUSING THE PLANET TO BECOME BARREN.

SOLAR WINDS CAN TRAVEL AT 1 MILLION MILES (1.6 MILLION KM) PER HOUR. THAT'S ABOUT 650 TIMES FASTER THAN A FIGHTER JET AT TOP SPEED!

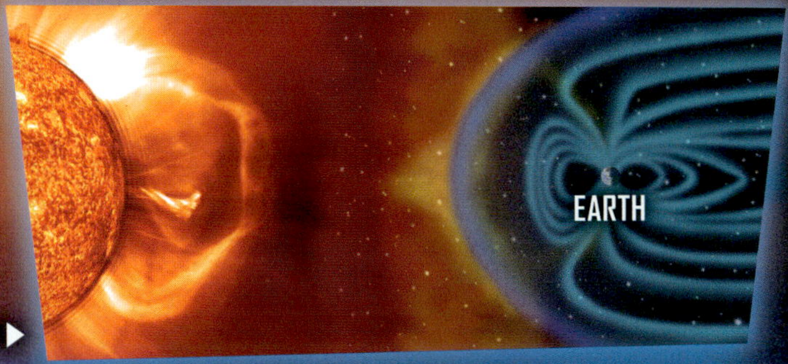

EARTH'S MAGNETIC SHIELD ▶

EARTH'S SHIELD

Fortunately for us, our planet is fairly protected from these solar winds because of its magnetic core. Earth's core is full of liquid metal. As Earth rotates around its axis, this metal magnetizes, creating a field around Earth. This magnetic field shields Earth and our atmosphere from the dangers of solar winds. However, our satellites, probes, and astronauts aren't as protected because they travel beyond Earth's atmosphere.

PIONEER 6, 7, 8, AND 9

The first satellites launched into space to study the sun were NASA's Pioneer probes. These satellites acted as a solar weather network, helping scientists predict solar storms that could impact communication. Just like TV meteorologists who tell us when it might rain or snow, the Pioneer satellites notified us of solar storms that were headed our way.

Pioneer 6 was launched on December 16, 1965. It was the first spacecraft to be tracked behind the sun, providing data that was previously unavailable. Along with its impressive data collection (such as new information on the corona), Pioneer 6 exceeded expectations. Originally **intended** to operate for just six months, Pioneer 6 managed to communicate data back to Earth for nearly 35 years after its launch.

INTERSTELLAR INFORMATION

IN 1986, THE PIONEER 7 PROBE PASSED HALLEY'S COMET AND GOT CLOSER THAN ANY OTHER SATELLITE OR PROBE BEFORE IT. PIONEER 7 STUDIED THE FAMOUS COMET'S TAIL AND ITS INTERACTION WITH SOLAR WINDS.

AFTER ITS LAUNCH, PIONEER 6 COLLECTED SOLAR WEATHER DATA THAT WAS SHARED WITH THE U.S. MILITARY, POWER COMPANIES, MAJOR AIRLINES, AND COMMUNICATION COMPANIES, SUCH AS PHONE PROVIDERS.

LET'S GO FOR A SPIN!

Some of the first of their kind launched into space, the Pioneer probes gave NASA the chance to try new ideas. One major advancement was the probes' ability to rotate themselves using a system called spin stabilization. This kept them steady and allowed them to resist other forces in space that might push them off course. It has now been used for other **deep space** probes, including Pioneer 10 on its mission to Jupiter and beyond.

THE HELIOS PROBES

As technology improved, scientists sought to get even closer to the sun. In 1974 and 1976, the U.S. and Germany teamed up to launch a pair of probes named Helios 1 and Helios 2. The Helios probes were equipped with several scientific instruments to analyze the sun's atmosphere, magnetic field, solar winds and more. The probes also had antennas that sent all this data back to Earth.

Another impressive feature was the heat **dispersal** system. This kept the probes rotating to avoid overheating as they reached temperatures upward of 700°F (371°C). The pair of probes were very successful, operating for about 10 years before they were shut down to avoid future interference with other probes. They provided data for scientists to study for years to come.

INTERSTELLAR INFORMATION

UP UNTIL 2018, HELIOS 2 HELD THE RECORD FOR THE CLOSEST APPROACH TO THE SUN BY A HUMAN-MADE OBJECT. SET IN 1976, HELIOS'S RECORD WAS 27 MILLION MILES (43.5 MILLION KM) FROM THE SUN'S SURFACE.

HELIOS 1 AND 2 WERE ALMOST IDENTICAL, BUT HELIOS 2 HAD IMPROVED SYSTEMS TO HELP IT SURVIVE LONGER IN SPACE.

A TEAM EFFORT

The Helios probes were the first satellites built outside the United States or Russia. NASA teamed up with the German Aeronautic and Space Research and Test Center, the space agency in what was West Germany. The probes were made in Munich, Germany. Once built, they launched in Cape Canaveral, Florida. When the probes sent information back to Earth, the scientists on the receiving end of those communications were located at telescopes in Australia, Germany, and the United States.

THE PARKER SOLAR PROBE

On August 12, 2018, NASA launched the Parker Solar Probe on a seven-year mission to fly closer to the sun than ever before. Parker's path will take it within 4 million miles (6.4 km) of the sun, which is seven times closer than any spacecraft before it.

The first sets of data generated by Parker were published on December 4, 2019, in the journal *Nature.* In this data set, scientists noted less dust close to the sun—something scientists thought might exist, but they had never observed. Finding a completely dust-free zone suggests the sun might have rings, just like Saturn! Scientists believe that the sun's dust-free zone is due to its intense heat, which converts nearby space dust to gas and leaves clear space in its place.

INTERSTELLAR INFORMATION

AS THE PARKER SOLAR PROBE ORBITS THE SUN, IT SPEEDS UP AND SLOWS DOWN. FOR ITS CLOSEST APPROACH, PARKER WILL REACH SPEEDS UP TO 430,000 MILES (692,018 KM) PER HOUR—THAT WOULD GET YOU FROM PHILADELPHIA, PENNSYLVANIA, TO WASHINGTON, DC, IN 1 SECOND!

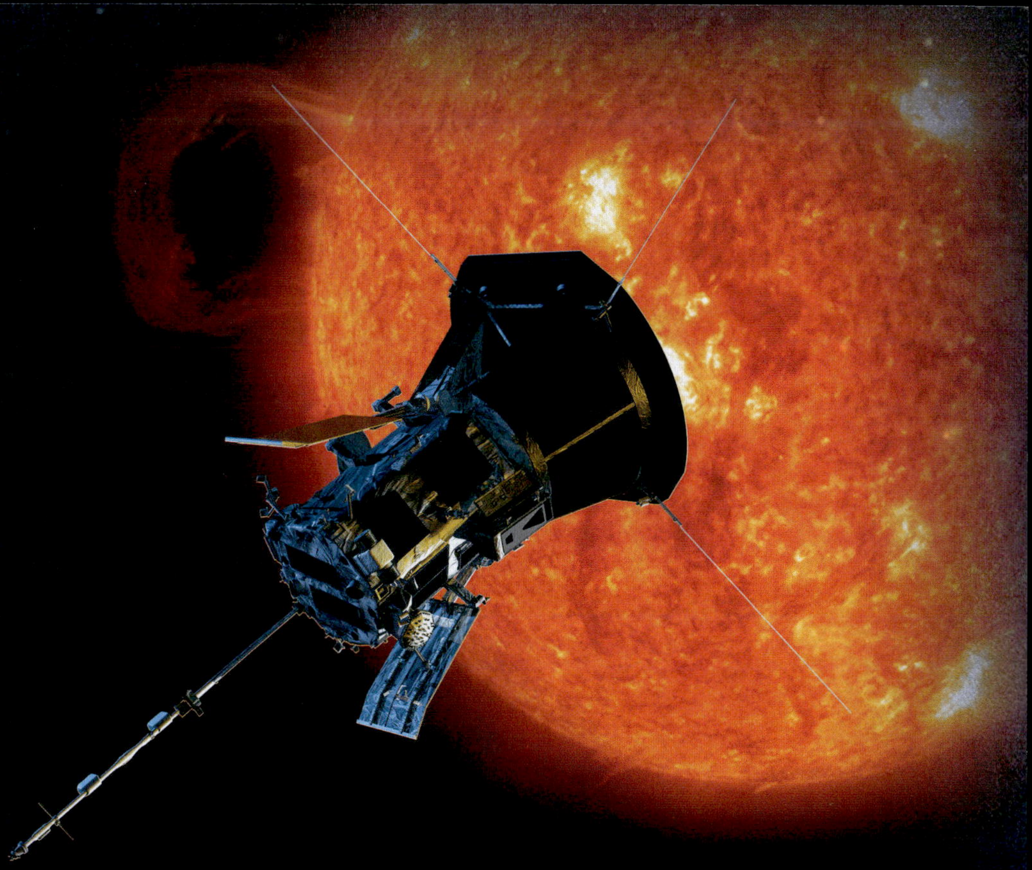

FOR ITS LAUNCH, THE PARKER SOLAR PROBE WAS MOUNTED TO AN ADDITIONAL ROCKET (ALSO CALLED A THIRD-STAGE ROCKET) TO **PROPEL** IT TOWARD THE SUN FASTER.

STANDING UP TO THE HEAT

To fly closer to the sun than ever, NASA engineers needed to design the Parker Solar Probe with heat tolerance in mind. Parker was expected to encounter temperatures near 2,500°F (1,371°C). Most things melt at that heat. Even lava erupting out of a volcano only reaches about 2,200°F (1,204°C)! Engineers designed a special carbon shield that could withstand the heat. They also coated parts of the shield in a white ceramic paint to reflect the intense sunlight away.

THE FUTURE OF SUN OBSERVATIONS

As electricity and technology take a bigger role in our lives, it's important for us to understand the sun and its solar winds, solar flares, and coronal mass ejections. We need to be ready for events such as the 1989 solar storm that plunged all of Quebec into darkness, or the strong solar flares in 2003 that caused severe damage to many orbiting satellites.

NASA's STEREO (Solar TErrestrial RElations Observatory) mission, launched in October 2006, is designed to observe the relationship between the sun and Earth—especially the flow of energy from the sun and how it impacts Earth. By studying and understanding the sun, with the goal of predicting its behavior, we can avoid the expensive and damaging effects from its radiation.

INTERSTELLAR INFORMATION

MORE COUNTRIES AND ORGANIZATIONS ARE BUILDING THEIR OWN SOLAR SATELLITES. FOR EXAMPLE, IN ITS FIRST MISSION TO STUDY THE SUN, THE INDIAN SPACE RESEARCH ORGANIZATION (ISRO) IS LAUNCHING THE ADITYA-1 PROBE IN 2020.

THIS SOLAR FLARE OCCURRED IN MAY 2013. IT WAS CAPTURED BY NASA'S SOLAR DYNAMICS OBSERVATORY.

UP CLOSE AND PERSONAL

One reason why we study our sun is because it's actually the only star we can learn from up close! The next closest star is Proxima Centauri, which is 4.24 light-years away. A light-year is how far light can travel in one Earth year. For comparison, the Helios 2 probe travels at about 149,000 miles (240,000 km) per hour. Even at that speed, it would still take 19,000 years to get there!

GLOSSARY

deep space: distant space and other planets of our solar system

disperse: to go or move in different directions

element: matter that is pure and has no other type of matter in it

intend: to plan or want to do something

intense: existing in an extreme degree

main sequence: a star that fuses hydrogen into helium in its core at a stable rate

perspective: a way of showing depth or distance in a drawing by making the objects that are far away smaller and making the objects that are closer to the viewer larger

propel: to push or drive something forward or in a particular direction

radiation: waves of energy

satellite: an object that circles Earth. Also, a scientific tool that circles Earth in order to collect and send information or aid in communication.

FOR MORE INFORMATION

BOOKS

Betts, Bruce, and Erica L. Colón. *Astronomy for Kids: How to Explore Outer Space with Binoculars, a Telescope, or Just Your Eyes.* Emeryville, CA: Rockridge Press, 2018.

Mosco, Rosemary, Jon Chad, and Luke Healy. *Solar System: Our Place in Space.* New York, NY: First Second, 2018.

Sabol, Stephanie, and Ted Hammond. *Where Is Our Solar System?* New York, NY: Penguin Workshop, 2018.

WEBSITES

NASA Kids Club
www.nasa.gov/kidsclub/index.html
Check out NASA's portal for kids with interactive games and other information.

NASA Space Place: Explore Earth and Space
spaceplace.nasa.gov/menu/sun/
What is a solar eclipse? Why do we get sunburned? Get answers to these questions and more on this NASA website for kids.

Sun Facts for Kids
www.planetsforkids.org/star-sun.html
Learn fun facts about history and science of the sun on this website.

Publisher's note to educators and parents: Our editors have carefully reviewed these websites to ensure that they are suitable for students. Many websites change frequently, however, and we cannot guarantee that a site's future contents will continue to meet our high standards of quality and educational value. Be advised that students should be closely supervised whenever they access the internet.

INDEX